Marx Joyce
Abbott Hardy Machiavelli Chesterton Cooper Emerson Austen
Defoe Melville Montaigne Haggard Hugo Grimm
Stoker Carroll Christie Molière Eliot
Wilde Maupassant Byron Schiller
Garnett Engels Smith Kafka
Goethe Einstein Fitzgerald Hawthorne Hall
Cotton Dostoyevsky Willis
Baum Kipling Doyle
Henry Nietzsche
Leslie Dumas Flaubert Turgenev Balzac
Stockton Vatsyayana Crane
Burroughs Verne
Curtis Tocqueville Gogol Vinci
Homer Widger Tolstoy Whitman Busch
Darwin Thoreau Twain
Potter Freud Zola Lawrence Scott Plato Harte
Kant Jowett Stevenson Dickens Burton Hesse
Andersen Cervantes
London Descartes Voltaire
Poe Aristotle Wells Cooke
Hale James Hastings
Bunner Shakespeare Irving
Richter Chekhov Chambers
Doré Dante da Shaw Benedict Alcott
Swift Wodehouse Pushkin
Newton

 tredition®

tredition was established in 2006 by Sandra Latusseck and Soenke Schulz. Based in Hamburg, Germany, tredition offers publishing solutions to authors and publishing houses, combined with world-wide distribution of printed and digital book content. tredition is uniquely positioned to enable authors and publishing houses to create books on their own terms and without conventional manufacturing risks.

For more information please visit: www.tredition.com

TREDITION CLASSICS

This book is part of the TREDITION CLASSICS series. The creators of this series are united by passion for literature and driven by the intention of making all public domain books available in printed format again - worldwide. Most TREDITION CLASSICS titles have been out of print and off the bookstore shelves for decades. At tredition we believe that a great book never goes out of style and that its value is eternal. Several mostly non-profit literature projects provide content to tredition. To support their good work, tredition donates a portion of the proceeds from each sold copy. As a reader of a TREDITION CLASSICS book, you support our mission to save many of the amazing works of world literature from oblivion. See all available books at www.tredition.com.

Project Gutenberg

The content for this book has been graciously provided by Project Gutenberg. Project Gutenberg is a non-profit organization founded by Michael Hart in 1971 at the University of Illinois. The mission of Project Gutenberg is simple: To encourage the creation and distribution of eBooks. Project Gutenberg is the first and largest collection of public domain eBooks.

The Scientific Evidences of Organic Evolution

George John Romanes

Imprint

This book is part of TREDITION CLASSICS

Author: George John Romanes
Cover design: Buchgut, Berlin – Germany

Publisher: tredition GmbH, Hamburg - Germany
ISBN: 978-3-8472-2987-2

www.tredition.com
www.tredition.de

NATURE SERIES.

THE SCIENTIFIC EVIDENCES OF ORGANIC EVOLUTION

BY

GEORGE J. ROMANES, M.A., LL.D., F.R.S.,
ZOOLOGICAL SECRETARY OF THE LINNEAN SOCIETY.

PREFACE.

Several months ago I published in the *Fortnightly Review* a lecture, which I had previously delivered at the Philosophical Institutions of Edinburgh and Birmingham, and which bore the above title. The late Mr. Darwin thought well of the epitome of his doctrine which the lecture presented, and urged me so strongly to republish it in a form which might admit of its being "spread broadcast over the land", that I promised him to do so. In fulfilment of this promise, therefore — which I now regard as more binding than ever — I reproduce the pg. viessay in the "Nature Series" with such additions and alterations as appear to me, on second thoughts, to be desirable. The only object of the essay is that which is expressed in the opening paragraph.

London,
June 1, 1882.

Since this little Essay was published, it has been suggested to me that, in its mode of presenting the arguments in favour of Evolution, there is a similarity to that which has been adopted by Mr. Herbert Spencer in the third part of his *Principles of Biology*. I should therefore like to state, that while such similarity is no doubt in part due to the similarity of subject-matter, I think, upon reading again, after an interval of ten years, his admirable presentation of the evidence it may also in part be due to unconscious memory. This applies particularly to the headings of the chapters, which I find to be almost identical with those previously used by Mr. Spencer.

G. J. R.

CONTENTS.

THE SCIENTIFIC EVIDENCES OF ORGANIC EVOLUTION.

Although it is generally recognised that the *Origin of Species* has produced an effect both on the science and the philosophy of our age which is without a parallel in the history of thought, admirers of Mr. Darwin's genius are frequently surprised at the ignorance of his work which is displayed by many persons who can scarcely be said to belong to the uncultured classes. The reason of this ignorance is no doubt partly due to the busy life which many pg. 2of our bread-winners are constrained to live; but it is also, I think, partly due to mere indolence. There are thousands of educated persons who, on coming home from their daily work, prefer reading literature of a less scientific character than that which is supplied by Mr. Darwin's works; and therefore it is that such persons feel these works to be-long to a category of books which is to them a very large one — the books, namely, which never are, but always to be, read. Under these circumstances I have thought it desirable to supply a short digest of the *Origin of Species*, which any man, of however busy a life, or of however indolent a disposition, may find both time and energy to follow.

With the general aim of the present abstract being thus under-stood, I shall start at the beginning of my subject by very briefly describing the theory of natural selection. It is a pg. 3matter of ob-servable fact that all plants and animals are perpetually engaged in what Mr. Darwin calls a "struggle for existence." That is to say, in every generation of every species a great many more individuals are born than can possibly survive; so that there is in consequence a perpetual battle for life going on among all the constituent individ-uals of any given generation. Now, in this struggle for existence, which individuals will be victorious and live? Assuredly those which are best fitted to live: the weakest and the least fitted to live will succumb and die, while the strongest and the best fitted to live will be triumphant and survive. Now it is this "survival of the fit-test" that Mr. Darwin calls "natural selection." Nature, so to speak, *selects* the best individuals out of each generation to live. And not only so, but as these favoured individuals transmit their favourable qualities to their offspring, according to the fixed pg. 4laws of heredity, it follows that the individuals composing each successive generation have a general tendency to be better suited to their surroundings than were their forefathers. And this follows, not merely because in every generation it is only the flower of the race that is allowed to breed, but also because if in any generation some new and beneficial qualities happen to appear as slight variations from the ancestral type, these will be seized upon by natural selection and added, by transmission in subsequent generations, to the previously existing type. Thus the best idea of the whole process will be gained by comparing it with the closely analogous process whereby gardeners and cattlebreeders create their wonderful productions; for just as these men, by always selecting their best individuals to breed from, slowly but continuously improve their stock, so Nature, by a similar process of selection, slowly but pg. 5continuously makes the various species of plants and animals better and better suited to the external conditions of their life.

Now, if this process of continuously adapting organisms to their environment takes place in nature at all, there is no reason why we should set any limits on the extent to which it is able to go up to the point at which a complete and perfect adaptation is achieved. Therefore we might suppose that all species would attain to this condition of perfect adjustment to their environment, and there remain fixed. And so undoubtedly they would, if the environment were itself unchanging. But forasmuch as the environment — or the sum total of the external conditions of life — of almost every organic type alters more or less from century to century (whether from astronomical, geological, and geographical changes, or from the immigrations and emigrations of other species living on contiguous pg. 6geographical areas), it follows that the process of natural selection need never reach a terminal phase. And forasmuch as natural selection may thus continue, *ad infinitum*, slowly to alter a specific type in adaptation to a gradually changing environment, if in any case the alteration thus effected is sufficient in amount to lead naturalists to denote the specific type by some different name, it follows that natural selection has transmuted one specific type into another. And so the process is supposed to go on over all the countless species of plants and animals simultaneously — the world of organic types being thus regarded as in a state of perpetual, though gradual, flux.

Such, then, is the theory of natural selection, or survival of the fittest; and the first thing we have to notice with regard to it is, that it offers to our acceptance a scientific explanation of the numberless cases of apparent design which we everywhere meet with in pg. 7organic nature. For all such cases of apparent design consist only in the *adaptation* which is shown by organisms to their environment, and it is obvious that the facts are covered by the theory of natural selection no less completely than they are covered by the theory of intelligent design. Perhaps it may be answered, — "The fact that these innumerable cases of adaptation may be accounted for by natural selection is no proof that they are not really due to intelligent design." And, in truth, this is an objection which is often urged by minds — even highly cultured minds — which have not been accustomed to scientific modes of thought. I have heard an eminent professor tell his class that the many instances of adaptation which Mr. Darwin discovered and described as occurring in orchids, seemed to him to tell more in favour of contrivance than in favour of natural causes; and another eminent professor once pg. 8wrote to me that although he had read the *Origin of Species* with care, he could see in it no evidence of natural selection which might not equally well be adduced in favour of intelligent design. But here we meet with a radical misconception of the whole logical attitude of science. For, be it observed, the exception *in limine* to the evidence which we are about to consider, does not question that natural selection *may* not be able to do all that Mr. Darwin ascribes to it: it merely objects to his interpretation of the facts, because it maintains that these facts might *equally well* be ascribed to intelligent design. And so undoubtedly they might, if we were all childish enough to rush into a supernatural explanation whenever a natural explanation is found sufficient to account for the facts. Once admit the glaringly illogical principle that we may assume the operation of higher causes where the operation of lower ones pg. 9is sufficient to explain the observed phenomena, and all our science and all our philosophy are scattered to the winds. For the law of logic which Sir William Hamilton called the law of parsimony — or the law which forbids us to assume the operation of higher causes when lower ones are found sufficient to explain the observed effects — this law constitutes the only logical barrier between science and superstition. For it is manifest that it is always possible to give a hypothetical explanation of any phenomenon whatever, by referring it immediately to the intelligence of some supernatural agent; so that the only difference between the logic of science and the logic of superstition consists in science recognising a validity in the law of parsimony which superstition disregards. Therefore I have no hesitation in saying that this way of looking at the evidence in favour of natural selection is not a scientific or a reasonable way of looking at it, but a pg. 10purely superstitious way. Let us take, for instance, as an illustration, a perfectly parallel case. When Kepler was unable to explain by any known causes the paths described by the planets, he resorted to a supernatural explanation, and supposed that every planet was guided in its movements by some presiding angel. But when Newton supplied a beautifully simple physical explanation, all persons with a scientific habit of mind at once abandoned the metaphysical explanation. Now, to be consistent, the above-mentioned professors, and all who think with them, ought still to adhere to Kepler's hypothesis in preference to Newton's explanation; for, excepting the law of parsimony, there is certainly no other logical objection to the statement that the movements of the planets afford as good evidence of the influence of guiding angels as they do of the influence of gravitation.

pg. 11So much, then, for the absurdly illogical position that, granting the evidence in favour of natural selection and supernatural design to be equal and parallel, we should hesitate for one moment in our choice. But, of course, if the evidence is supposed *not* to be equal and parallel — *i.e.*, if it is supposed that the theory of natural relation is not so competent a theory to explain the facts of adaptation as is that of intelligent design — then the objection is no longer the one that we are considering. It is quite another objection, and one which is not *primâ facie* absurd; it requires to be met by examining how far the theory of natural selection *is* able to explain the facts. Let us state the problem clearly.

Innumerable cases of adaptation of organisms to their environment are the observed facts for which an explanation is required. To supply this explanation two, and only two, hypotheses are in the field. Of these pg. 12two hypotheses one is, intelligent design manifested in creation; and the other is, natural selection manifested during the countless ages of the past. Now it would be proof positive of intelligent design if it could be shown that all species of plants and animals were *created*—that is *suddenly* introduced into the complex conditions of their life; for it is quite inconceivable that any cause other than intelligence could be competent to adapt an organism to its environment *suddenly*. On the other hand, it would be proof presumptive of natural selection if it could be shown that one species becomes slowly transmuted into another—*i.e.*, that one set of adaptations may be gradually transformed into another set of adaptations according as changing circumstances require. This would be proof presumptive of natural selection, because it would then become amply probable that natural selection might have brought about many, or most, of the cases of pg. 13adaptations which we see; and if so, the law of parsimony excludes the rival hypothesis of intelligent design. Thus the whole question as between natural selection and supernatural design resolves itself into this—Were all the species of plants and animals separately created, or were they slowly evolved? For if they were specially created, the evidence of supernatural design remains unrefuted and irrefutable; whereas if they were slowly evolved, that evidence has been utterly and for ever destroyed. The doctrine of natural selection therefore depends for its validity on the doctrine of organic evolution; for if once the fact of organic evolution were established, no one would dispute that much of the adaptation was probably effected by natural selection. *How* much we cannot say—probably never shall be able to say; for even Mr. Darwin himself does not doubt that other causes besides that of natural selection have assisted in the modifyingpg. 14 of specific types. For the sake of simplicity, however, I shall not go into this subject; but shall always speak of natural selection as the only cause of organic evolution. Let us, then, weigh the evidence in favour of organic evolution. If we find it wanting, we need have no complaints to make of natural theologians of to-day; but if we find it to be full measure, shaken together and running over, we ought to maintain that natural theologians can no longer adhere to the arguments of such writers as Paley, Bell, and Chalmers, without deliberately violating the only logical principle which separates science from fetishism.

To avoid misapprehension, however, I may here add that while Mr. Darwin's theory is thus in plain and direct contradiction to the theory of design, or system of teleology, as presented by the school of writers which I have named, I hold that Mr. Darwin's theory has no point of logical contactpg. 15 with the theory of design in the larger sense, that behind all secondary causes of a physical kind, there is a primary cause of a mental kind. Therefore throughout this essay I refer to design in the sense understood by the narrower forms of teleology, or as an *immediate* cause of the observed phenomena. Whether or not there is an *ultimate* cause of a psychical kind pervading all nature, a *causa causarum* which is the final *raison d'être* of the cosmos, this is another question which, as I have said, I take to present no point of logical contact with Mr. Darwin's theory, or, I may add, with any of the methods and results of natural science. The only position, therefore, which I here desire to render plain is that, if the doctrine of evolution is seen to be established by sufficient evidence, and therefore the causes which it sets forth are recognised as adequate to furnish a scientific explanation of the results observed, then the pg. 16 facts of organic nature necessarily fall into the same logical category, with reference to any question of design, as that of all or any other series of facts in the physical universe.

This being understood, I shall now proceed to render an epitome of the evidence in favour of organic evolution, and I shall do so by classifying the arguments in a way tending to show their distinct or independent character, and therefore calculated to display the additional force which they acquire from their cumulative nature.

pg. 17

I.

THE ARGUMENT FROM CLASSIFICATION.

I shall first take the argument from classification. Naturalists find that all species of plants and animals present among themselves structural affinities. According as these structural affinities are more or less pronounced, the various species are classified under genera, orders, families, classes, sub-kingdoms, and kingdoms. Now in such a classification it is found impossible to place all the species in a linear series, according to the grade of their organization. For instance, we cannot say that a wolf is more highly organized than a fox or a jackal; we can only say that the pg. 18specific points wherein it differs from these animals are without significance as proving the one type to be more highly organized than the others. But of course in many cases, and especially in the cases of the larger divisions, it is often possible to say — The members in this division are more highly organized than are the members in that division. Our system of classification therefore may be likened to a tree, in which a short trunk may be taken to represent the lowest organisms which cannot properly be termed either plants or animals. This short trunk soon separates into two large trunks, one of which represents the vegetable and the other the animal kingdom. Each of these trunks then gives off large branches signifying classes, and these give off smaller, but more numerous branches, signifying families, which ramify again into orders, genera, and finally into the leaves, which may be taken to represent species. Now, pg. 19in such a representative tree of life, the height of any branch from the ground may be taken to indicate the grade of organization which the leaves, or species, present; so that, if we picture to ourselves such a tree, we will understand that while there is a general advance of organization from below upwards, there are numberless slight variations in this respect between leaves growing even on the same branch; but in a still greater number of cases, leaves growing on the same branch are growing on the same level — that is, although they represent different species, it cannot be said that one is more highly organized than the other. Now, this tree-like arrangement of specific organisms in nature is an arrangement for which Mr. Darwin is not responsible. I mean that the framing of this natural classification has been the work of naturalists for centuries past; and although they did not know what they were doing, it is now evident to evolutionists that they pg. 20were tracing the lines of genetic relationship. For, be it observed, a scientific or natural classification differs very much from a popular or hap-hazard classification, and the difference consists in this, that while a popular classification is framed with exclusive reference to the external appearance of organisms, a scientific classification is made with reference to the whole structure. A whale, for instance, is often thought to be a fish, because it resembles a fish in form and habits; whereas dissection shows that it is beyond all comparison more unlike a fish than it is like a horse or a man. This is, of course, an extreme case; but it was cases such as this that first led naturalists to see that there are resemblances between organisms much more deep and important than appear upon the surface; and consequently, that if a natural classification was possible at all, it must be made with reference to these deeper resemblances. Of course, it took pg. 21time to perceive this distinction between fundamental and superficial resemblances. I remember once reading a very comical disquisition in one of Buffon's works on the question as to whether or not a crocodile was to be classified as an insect; and the instructive feature in the disquisition was this, that although a crocodile differs from an insect as regards every conceivable particular of its internal anatomy, no allusion at all is made to this fact, while the whole discussion is made to turn on the hardness of the external casing of a crocodile resembling the hardness of the external casing of a beetle; and when at last Buffon decides that, on the whole, a crocodile had better not be classified as an insect, the only reason given is, that as a crocodile is so very large an animal, it would make "altogether too terrible an insect."

But now, when at last it came to be recognised that internal anatomy rather than externalpg. 22 appearance was to be taken as a guide to classification, the question was, What features in the internal anatomy are to take precedence over the other features? And this question it was not hard to answer. A porpoise, for instance, has a large number of teeth, and in this feature resembles most fish, while it differs from all mammals. But it also gives suck to its young, and in this feature it differs from all fish, while it resembles all mammals. Now, looking at those two features alone, should we say that a porpoise ought to be classed as a fish or as a mammal? Assuredly as a mammal, and for this reason: The number of teeth is a very variable feature both in fish and in mammals, whereas the giving of suck is an invariable feature among mammals, and occurs nowhere else in the animal kingdom. This, of course, is purposely chosen as a very simple illustration; but it exemplifies the general fact that thepg. 23 guiding principle of scientific classification is the comparing of organism with organism, with the view of seeing which of the constituent organs are of the most invariable occurrence, and therefore of the most typical signification.

Now, since the days of Linnæus this principle has been carefully followed, and it is by its aid that the tree-like system of classification has been established. No one, even long before Darwin's days, ever dreamed of doubting that this system is in reality, what it always has been in name, a *natural* system. What, then, is the inference we are to draw from it? An evolutionist answers, that it is just such a system as his theory of descent would lead him to expect as a natural system. For this tree-like system is as clear an expression as anything could be of the fact that all species are bound together by the ties of genetic relationship. If all species were separately created, it is almost incrediblepg. 24 that we should everywhere observe this progressive shading off of characters common to larger groups, into more and more specialized characters distinctive only of smaller and smaller groups. At any rate, to say the least, the law of parsimony forbids us to ascribe such effects to a supernatural cause, acting in so whimsical a manner, when the effects are precisely what we should expect to follow from the action of a highly probable natural cause. The classification of animal forms, indeed, as Darwin, Lyell, and Hæckel have pointed out, strongly resembles the classification of languages. In the case of languages, as in the case of species, we have genetic affinities strongly marked; so that it is possible to some extent to construct a language-tree, the branches of which shall indicate, in a diagrammatic form, the progressive divergence of a large group of languages from a common stock. For instance, Latin may be regarded as a fossilpg. 25 language, which has given rise, by way of genetic descent, to a group of living languages—Italian, Spanish, French, and, to a large extent, English. Now what should we think of a philologist who should maintain that English, French, Spanish, and Italian were all specially created languages — or languages separately constructed by the Deity, and by as many separate acts of inspiration communicated to these several nations — and that their resemblance to the fossil form, Latin, is to be attributed to special design? Yet the evidence of the natural transmutation of species, is, in one respect, much stronger than that of the natural transmutation of languages — in respect, namely, of there being a vastly greater number of cases all bearing testimony to the fact of genetic relationship.

pg. 26

II.

THE ARGUMENT FROM MORPHOLOGY OR STRUCTURE.

I now pass to another line of argument. The theory of evolution by natural selection supposes that hereditary characters admit of being slowly modified wherever their modification will render an organism better suited to a change in its conditions of life. Let us, then, observe the evidence we have of such adaptive modifications of structure, in cases where the need of such modification is apparent. For the sake of clearness, I shall begin by again taking the case of the whales and porpoises. The theory of pg. 27evolution infers, from the whole structure of these animals, that their progenitors must have been terrestrial quadrupeds of some kind, which became aquatic in their habits. Now the change in the conditions of their life thus brought about would render desirable great modifications of structure. These changes would, in the first instance, begin to affect the least typical—that is, the least strongly inherited structures— such as the skin, claws, and teeth, &c. But as time went on, the adaptation would begin to extend to the more typical structures, until the shape of the body began to be affected by the bones and muscles required for terrestrial locomotion becoming better adapted for aquatic locomotion, and the whole outline of the animal more fish-like in shape. This is the stage which we actually observe in the seals, where the hind legs, although retaining all their typical bones, have become shortened up almost to pg. 28rudiments, and directed backwards, so as to be of no use for walking, but serving to complete the fish-like taper of the body. But in the whales the modification has gone even further than this, so that the hind legs have ceased to be apparent externally, and are only represented internally by remnants so rudimentary that it is impossible to make out with certainty the homologies of the bones; moreover, the head and the whole body have become completely fish-like in shape. But profound as these changes are, they only affect those parts of the organism which it was for the benefit of the organism to have altered, so that it might be adapted to an aquatic mode of existence. Thus the arm, which is used as a fin, still retains the bones of the shoulder, fore-arm, wrist, and fingers, although they are all inclosed in a fin-shaped sack, so as to render them quite useless for any other purpose pg. 29than swimming. Similarly, the head, although it so closely resembles the head of a fish in shape, still retains the bones of the mammalian skull in their proper anatomical relation to one another, but modified in form so as to offer the least possible amount of resistance to the water. In short it may be said that all the modifications have been effected with the least possible divergence from the typical mammalian type, which is compatible with securing so perfect an adaptation to a purely aquatic mode of life.

Now I have chosen the case of the whale and porpoise group because they offer so extreme an example of profound modification of structure in adaptation to changed conditions of life. But the same thing may be seen in hundreds and hundreds of other cases. For instance, to confine our attention to the arm, not only is the limb modified in the whale for swimming, but in another mammal—the bat—it is modified forpg. 30 flying, by having the fingers enormously elongated and overspread with a membranous web. In birds, again, the arm is modified for flight in a wholly different way—the fingers here being very short and all run together, and the chief expanse of the wing being composed of the shoulder and fore-arm. In frogs and lizards, again, we find hands more like our own; but in an extinct species of flying reptile the modification was extreme, the wing having been formed by a prodigious elongation of the fifth finger, and a membrane spread over it and the rest of the hand. Lastly, in serpents the hand and arm have disappeared altogether.

Thus, even if we confine our attention to a single structure, how wonderful are the modifications which it is seen to undergo, although never losing its typical character! How are we to explain this? By design manifested in special creation, or by descent with adaptivepg. 31 modification? If it is said by design manifested in special creation, we must suppose that the Deity formed an archetypal plan of certain structures, and that He determined to adhere to this plan through all the modifications which those structures exhibit. Now the difficulties in the way of this supposition are prodigious, if not quite insurmountable. In the first place, why is it that some structures are selected as typical and not others? Why should the vertebral skeleton, for instance, be tortured into every conceivable variety of modification in order to make it serviceable for as great a variety of functions; while another structure, such as the eye, is made in different sub-kingdoms on fundamentally different plans, notwithstanding that it has throughout to perform the same function? Will any one have the hardihood to assert that in the case of the skeleton the Deity has endeavoured to show His *ingenuity* by the manifold functions to which He has made the same structure pg. 32subservient; while in the case of the eye He has endeavoured to show his *resources* by the manifold structures which He has to subserve the same function? If so, it appears to me a most unfortunate circumstance, that throughout both the vegetable and animal kingdoms, all cases which can be pointed to as showing ingenious adaptation of the same typical structure to the performance of widely different functions, are cases which come within the limits of the same natural group of plants and animals, and therefore admit of being equally well explained by descent from a common ancestry; while all cases of widely different structures performing the same function are to be found in different groups of plants or animals, and are therefore suggestive of independent variations arising in the different lines of hereditary descent.

To take a specific illustration. The octopuspg. 33 or devil-fish belongs to a widely different class of animals from a true fish, and yet its eye, in general appearance, looks wonderfully like the eye of a true fish. Now, Mr. Mivart pointed to this fact as a great difficulty in the way of the theory of evolution by natural selection, because it must clearly be a most improbable thing that so complicated a structure as the eye of a fish should happen to be arrived at through each of two totally different lines of descent. And this difficulty would, indeed, be almost fatal to the theory of evolution by natural selection, if the apparent similarity were a real one. Unfortunately for the objection, however, Mr. Darwin clearly showed, in his reply, that in no one anatomical feature of typical importance do the two structures resemble one another; so that in point of fact the two organs do not resemble one another in any particular further than it is necessary that they should, if both are to serve as organs ofpg. 34 sight. But now, suppose that this had not been the case, and that the two structures, besides presenting the necessary superficial resemblance, had also presented an anatomical resemblance; with what tremendous force might it have then been urged,—"Your hypothesis of hereditary descent with progressive modification being here excluded, by the fact that the animals compared belong to two widely different branches of the tree of life, how are we to explain the identity of type manifested by these two complicated organs of vision? The only hypothesis open to us is intelligent adherence to an ideal type." But as this cannot now be urged in any one case throughout the whole organic world, we may, on the other hand, present it as a most significant fact, that, while within the limits of the same large branch of the tree of life we constantly find the same typical structures modified so as to perform very different functions, we never find any vestige ofpg. 35 these particular types of structure in other large divisions of that tree. In other words, we never find typical structures appearing except in cases where their presence may be explained by the hypothesis of hereditary descent; while in thousands of such cases we find these structures undergoing every conceivable variety of adaptive modification.

Consequently, special creationists must fall back upon another position and say,—"Well, but it may have pleased the Deity to form a certain number of ideal types, and never to allow the structures occurring in the one type to appear in any of the others." I answer, undoubtedly it may have done so; but if it did, it is a most unfortunate thing for your theory; for the fact implies that the Deity has planned His types in such a way as to suggest the counter-theory of descent. For instance, it would seem to me a most capricious thing inpg. 36 the Deity to make the eyes of an innumerable number of fish on exactly the same ideal type, and then to make the eye of the octopus so exactly like these other eyes in superficial appearance as to deceive so accomplished a naturalist as Mr. Mivart, and yet to take scrupulous care that in no one ideal particular should this solitary eye resemble all the host of other eyes. However, adopting for the sake of argument this gigantic assumption, let us suppose that God laid down these arbitrary rules for His own guidance in creation, and let us see to what it leads. If, as is assumed, the Deity formed a certain number of ideal types, and determined that on no account should He allow any part of one type to appear in any part of another, surely we should expect that within the limits of the same type the same typical structures should always be present. Thus, remember what desperate efforts, so topg. 37 speak, there have been made to maintain the uniformity of type in the case of the arm, and should we not expect that in other and similar cases similar efforts should be made? Yet we repeatedly find that this is not the case. Even in the whale, as we have seen, the hind-limbs are not apparent; and it is impossible to see in what respect the hind-limbs are of any less ideal value than the fore-limbs, which, as we have also seen, are so carefully preserved in nearly all vertebrated animals except the snakes, where again we meet in this particular with a sudden and sublime indifference to the maintenance of a typical structure. Now I say that if the theory of ideal types is true, we have in these facts evidence of the most unreasonable inconsistency; for no explanation can be assigned why so much care should have been taken to maintain the type in some cases, while such reckless indifference should have been displayedpg. 38 towards maintaining it in others. But the theory of descent with continued adaptive modification fully explains all the known cases; for in every case the degree of divergence from the typical structure which an organism presents corresponds with the length of time during which the divergence has been going on. Thus we scarcely ever meet with any great departure from the typical form—such as the absence of limbs—without some of the other organs in the body being so far modified as of themselves to indicate, on the supposition of descent with modification, that the animal or plant must have been subject to the modifying influences for a long series of generations. And this combined testimony of a number of organs in the same organism is what the theory of descent would lead us to expect, while the rival theory of design can offer no explanation of the fact, that when one organ shows a conspicuous pg. 39departure from the supposed ideal type, some of the other organs in the same organism should tend to keep it company by doing likewise.[1]

I will now briefly touch on another branch of the argument from morphology—the argument, namely, from rudimentary structures.

Throughout the animal and vegetable kingdoms we constantly meet with organs which are the dwarfed and useless representatives of organs which, in other and allied kinds of animals and plants, are of large size and functional utility. Thus, for instance, the unborn whale has rudimentary teeth, which are never destined to cut the gums; and we all know that our own rudimentary tail is of no practical service. Now, rudimentary organs of this kind are of suchpg. 40 common occurrence, that almost every species presents one or more of them. The question, therefore, is—How are they to be accounted for? Of course the theory of descent with adaptive modification has a delightfully simple answer to supply, viz., that when, from changed conditions of life, an organ which was previously useful becomes useless, natural selection, combined with disuse and so-called economy of growth, will cause it to dwindle till it becomes a rudiment. On the other hand, the theory of special creation can only maintain that these rudiments are formed for the sake of adhering to an ideal type. Now, here again the former theory is triumphant over the latter; for, without waiting to dispute the wisdom of making dwarfed and useless structures merely for the whimsical motive assigned, surely if so extraordinary a method is adopted in so many cases, we should expect that in consistency it would be adopted in all cases. Thispg. 41 reasonable expectation, however, is far from being realised. In numberless cases, such as that of the fore-limbs of serpents, no vestige of a rudiment is present. But the vacillating policy in the matter of rudiments does not end here; for it is shown, if possible, in a more aggravated form where, within the limits of the same natural group of organisms, a rudiment is sometimes present and sometimes absent. For instance, to take again the case of limbs, in nearly all the numerous species of snakes there are no vestiges of limbs at all; but in the python we find beneath the skin very tiny rudiments of the hind limbs. Now, is it a worthy conception of Deity that, while neglecting to maintain His unity of ideal in the case of nearly all the numerous species of snakes, He should have added a tiny rudiment in the case of the python, and even in that case should have maintained His ideal type very inefficiently, inasmuchpg. 42 as only two limbs instead of four are represented? Or, again, take the case of the limb in other animals. Five toes seem to constitute the ideal type, notwithstanding that in numberless cases this ideal fails in its structural expression. Now, in the case of the horse, one toe appears to have become developed at the expense of the others; for the so-called knee of the horse is really the wrist or ankle, and the so-called shank the middle toe or finger very much enlarged. But on each side of this enlarged toe there are, beneath the skin, rudimentary bones of two other toes—the so-called splint-bones. So far good, but three toes are not five; so special creationists must suppose that while in this case the

Deity has, so to speak, struggled to maintain the uniformity of His ideal, His efforts have nevertheless conspicuously failed. How much less strained is the scientific interpretation; for I may mention that in this particularpg. 43 case, besides the general inference that rudiments point us to a remote ancestry, we have direct palæontological evidence that there have been a whole series of extinct horse-like animals, that began low down in the geological strata with five toes (on the fore-feet, one being rudimentary), which afterwards became reduced to four and then to three; after which the two lateral toes began to become rudimentary, as we now see them in oxen, and later on still more so. Lastly, as we come nearer to recent times, we find fossils of the existing horse, with the lateral toes shortened up to the condition of splint-bones. Thus we have some half-dozen different genera of horse, all standing in a linear series in time as in structure, between the earliest representative with the typical number of five toes, and the existing very aberrant form with only one toe.

It is sometimes said that a striking pg. 44corroboration of a scientific theory is furnished when it enables us correctly to *predict* discoveries. Such a corroboration is afforded in this instance; for Professor Huxley, speaking in 1870, said, "If the expectation raised by the splints of the horses that, in some ancestor of the horses, these splints would be found to be complete digits, has been verified, we are furnished with very strong reasons for looking for a no less complete verification that the three-toed *plagiolophus*-like 'avus' of the horse must have had a five-toed 'atavus' at some earlier period. No such five-toed 'atavus,' however, has yet made its appearance." But since then the "atavus" has made its appearance, if not with five complete toes, at least with four complete and one rudimentary; and any day we may hear that Professor Marsh has found in still earlier strata a more primitive form with all five toes complete.pg. 45

I have no space to go into the evidence of similar "missing links" which have been recently supplied by palæontological researches in the case of several other groups of animals; but their consideration seems to me quite to justify a more recent utterance of Professor Huxley, who, in 1878, wrote in the *Encyclopædia Britannica:* "On the evidence of palæontology, the evolution of many existing forms of animal life from their predecessors is no longer an hypothesis, but an historical fact; it is only the nature of the physiological factors to which that evolution is due which is still open to discussion."

[1] This consideration is, I believe, original. Several exceptions to its validity might be adduced, but as a general principle it certainly holds good.

pg. 46

III.

THE ARGUMENT FROM GEOLOGY.

But this allusion to fossils leads me to the next division of my subject—the argument from geology. It is not, however, necessary to say much on this head, for the simple reason that the whole body of geological evidence is for the most part of one kind, which although of a very massive, is of a very simple character. That is to say, apart from the increasingly numerous cases, such as the one just mentioned, which geology supplies of extinct "intermediate links" between *particular* species now living, the great weight of the geological evidence consists pg. 47in the *general* fact, that of all the thousands of specific forms of life which palæontology reveals to us as having lived on this planet in times past, there is no instance of a highly organised form occurring low down in the geological series.[1] On the contrary, there is the best evidence to show that since the first dawn of life in the occurrence of the simplest organisms, until the meridian splendour of life as now we see it, gradual advance from the general to the special—from the low to the high, from the few and simple to the many and complex—has been the law of organic nature. And of course it is needless to say that this is precisely the law to which the process of descent with adaptive modification would of necessity give rise.

[1] Some of the lower vertebrata (Elasmobranch and Ganoid fishes) occur, indeed, in early strata (upper Silurian); but still far from the earliest in which some of the invertebrata are found. The general statement in the text applies chiefly to the more highly organised forms of the vertebrate series.

pg. 48

IV.

THE ARGUMENT FROM GEOGRAPHICAL DISTRIBUTION.

The argument from geology is the argument from the distribution of species in time. I will, therefore, next take the argument from the distribution of species in space—that is, the present geographical distribution of plants and animals. It is easy to see that this must be a most important argument, if we reflect that as the theory of descent with adaptive modification implies slow and gradual change of one species into another, and a still more slow and gradual change of one genus, family, or order pg. 49into another genus, family, or order, we should expect on this theory that the organic types living on any given geographical area should be found to resemble or to differ from organic types living elsewhere, according as the area is connected or disconnected with other geographical areas. And this we find to be the case, as abundant evidence proves. For, to quote from Mr. Darwin, "barriers of any kind, or obstacles to free migration, are related in a close and important manner to the differences between the productions of various regions. We see this in the great difference in nearly all the terrestrial productions of the New and Old Worlds, excepting in the northern parts, where the land almost joins.... We see the same fact in the great difference between the inhabitants of Australia, Africa, and South America under the same latitude, for these countries are almost as much isolated from one another as possible. On pg. 50each continent, also, we see the same fact; for on the opposite sides of lofty and continuous mountain ranges, of great deserts, and even of large rivers, we find different productions; though as mountain chains, deserts, &c., are not so impassable, or likely to have endured so long as the ocean-separated continents, the differences are very inferior in degree to those characteristic of distinct continents." That is to say, the differences are usually confined to species and genera, whereas in the case of continents the differences extend to orders. Similarly in marine productions the same laws prevail—the species on the different sides of the American continent, for instance, being very distinct. Now, this law cannot be explained by any reasonable argument from design.

And still stronger does the present argument become when we look to the fossil species contained on different continents; for these fossil speciespg. 51 invariably present the same characteristic stamp as the living species now flourishing on the same continents. Thus, in America we find fossils all presenting the characteristically American types of animals, in Australia the characteristically Australian types, and so on. That is to say, on every continent the dead species resemble the living species, as we may expect that they should, if they are all bound together by the ties of hereditary descent; while, if different continents are compared, the fossil species are as unlike as we have seen the living species to be.

Turning next to the case of oceanic islands, situated at some distance from a continent. In these cases the plants and animals found on the island, though very often differing from all other plants and animals in the world as regards their specific type, nevertheless in generic type resemble the plants and animals of the neighbouring continent. The inference clearly is, thatpg. 52 the island has been stocked from the continent with these types—either by winds, currents, floating trees, or numerous other modes of transport—and that, after settling in the island, some of these imported types have retained their specific characters, while others have varied so as to become specific types peculiar to that island. The Galapagos Archipelago islands are particularly instructive in this connection; for while the whole group of islands lies at a distance of over five hundred miles from the shores of South America, the constituent islands are separated from one another by straits varying from twenty to thirty miles. Now, to quote from Darwin, "Each separate island of the Galapagos Archipelago is tenanted, and the fact is a marvellous one, by many distinct species; but these species are related to each other in a very much closer manner than to the inhabitants of the American continent." That is to say,pg. 53 the American continent being some fifteen times the distance from these islands that they are from one another, emigration to them from the continent is of much more rare occurrence than emigration from one island to another; and therefore, as more time for variation is thus allowed, while the differences between the inhabitants of island and island are only specific, the differences between the inhabitants of the islands as a group and the inhabitants of the American continent are very often generic. I may mention, in passing, that it was upon discovering these relations in the case of the Galapagos Archipelago, and pondering upon them as "marvellous facts," that Mr. Darwin was first led to entertain the idea that the doctrine of descent might be the grand truth for which the science of the nineteenth century was waiting.

The evidence from oceanic islands, however,pg. 54 is not yet exhausted; for in no part of the world is there an oceanic island more than a certain distance from a mainland in which any species of the large class of frogs, toads, and newts is to be found. Why is this? Simply because these animals, and their spawn, are quickly killed by contact with sea-water; and therefore frogs, toads, and newts have never been able to reach oceanic islands in a living state. Similarly in all oceanic islands situated more than three hundred miles from land, no species of the whole class of mammals is to be found, excepting species of the only order of mammals which can fly, viz., bats. And, as if to make the case still stronger, these forlornly created species of bats sometimes differ from all other bats in the world. But can we, as reasonable men, suppose that the Deity has chosen, without any apparent reason, never to create any frog, toad, newt, or mammal on any oceanic island,pg. 55 save only such species as are able to fly? Or, if we go so far as to say, — "There may have been some hidden reason why batrachians and quadrupeds should not have been created on oceanic islands," I will adduce another very remarkable fact, viz., that on some of these islands there occur species of plants, the seeds of which are provided with numerous hooks adapted to catch the hair of moving quadrupeds, and so to become disseminated. But, as we have just seen, there are no quadrupeds in these islands to meet this case of adaptation; so that special creationists must resort to the almost impious hypothesis, that in these cases the Deity only carried out half His plan, in that while He made an elaborate provision for plants which depended for its efficiency on the presence of quadrupeds, He nevertheless, after all, neglected to place the quadrupeds in the same islands as the plants! Now, I submit that such abortive pg. 56attempts at adaptation bring the thesis of the special creationists to a *reductio ad absurdum*; so that the only possible explanation before us is, that while the seeds of these plants were able to float to the islands, the quadrupeds were not able to swim.

Perhaps in sheer desperation, however, the special creationists will try to take refuge in the assumption that oceanic islands differ from continents in not having been the scenes of creative power, and have therefore depended on immigration for their inhabitants. But here again there is no standing-room; for we have already seen that oceanic islands are particularly rich in peculiar species which occur nowhere else in the world; so that, as a matter of fact, if the special creation theory is true, we must conclude that oceanic islands have been the theatres of extraordinary creative activity; although an exception has always been carefully made to the detrimentpg. 57 of frogs, toads, newts, and mammals, save only such as are able to fly.

If space permitted, I might adduce several other highly instructive facts in this argument from geographical distribution; but I will content myself with mentioning only one other. When Mr. Wallace was at the Malay Archipelago, he observed that the quadrupeds inhabiting the various islands belonged to the same or to closely allied species. But he also observed that all the quadrupeds inhabiting the islands lying on one side of an imaginary sinuous line, differed widely from the quadrupeds inhabiting the islands lying on the other side of that line. Now, soundings showed that in exact correspondence with this imaginary sinuous line the sea was much deeper than in any other part of the Archipelago. Consequently, how beautiful is the explanation. We have only to suppose that at some previous time the sea bottom waspg. 58 raised sufficiently to unite all the islands on each side of the deep water into two great tracts of land, separated from one another by the deep strait of water. Each of these great tracts of land would then have had their own distinctive kinds of quadrupeds—just as the American quadrupeds are now distinct from the European; for the comparatively narrow strait between the then Malay continents would have offered as effectual a barrier to the migration of quadrupeds as does the Atlantic Ocean at the present day. Hence, when all the land slowly subsided so as to leave only its mountain chains and table lands standing above the surface in the form of islands, we now have the state of things which Mr. Wallace describes—viz., two large groups of islands with the quadrupeds on the one group differing widely from the quadrupeds on the other, while within the limits of the same group the quadrupeds inhabitingpg. 59 different islands all belong to the same or to closely allied species. On this highly interesting subject Darwin writes, "I have not as yet had time to follow up this subject in all quarters of the globe; but as far as I have gone the relation holds good. For instance, Britain is separated by a shallow channel from Europe, and the mammals are the same on both sides, and so it is with all the islands near the shores of America. The West Indian islands, on the other hand, stand on a deeply submerged bank nearly 1,000 fathoms in depth, and here we find American forms, but the species, and even the genera, are distinct. As the amount of modification which animals of all kinds undergo partly depends on lapse of time, and as the islands which are separated from each other or from the mainland by shallow channels are more likely to have been continuously united within a recent period than the islands separated bypg. 60 deeper channels, we can understand how it is that a relation exists between the depth of the sea separating two mammalian faunas, and the degree of their affinity — a relation which is quite inexplicable on the theory of independent acts of creation."

So much, then, for the argument from geographical distribution — the many facts of crucial importance which it affords almost resembling so many experiments devised by Nature to prove the falsity of the special creation hypothesis. For now, let it in conclusion be observed, that there is no *physiological* reason why animals and plants of the different characters observed should inhabit different continents, islands, seas, and so forth. As Darwin observes, "there is hardly a climate or condition in the Old World which cannot be paralleled in the New ... and yet how widely different are their living productions." And that it is not the suitability of organismspg. 61 to the areas which they inhabit which has determined their creation upon those areas, is conclusively proved by the effects of the artificial transportation of species by man. For in such cases it frequently happens that the imported species thrives quite as well in its new as in its old home, and indeed often supplants the native species. As the Maoris say, — "As the white man's rat has driven away the native rat, so the European fly has driven away our fly, so the clover kills our fern, and so will the Maori himself disappear before the white man."

Upon the whole then we are driven to the conclusion, that if the special creation theory is true, the various plants and animals have not been placed in the various habitats which they occupy with any reference to the suitability of these habitats to the organisations of these particular plants and animals. Sopg. 62 that, considering all the evidence under the head of geographical distribution, I think we are driven to the yet further conclusion, that if the special creation theory is true, the only principle which appears to have been consistently followed in the geographical deposition of species, is the principle of so depositing them as in all cases to make it appear that the supposition of their having been thus deposited is not merely a highly dubious one, but one which, on the face of it, is conspicuously absurd.

pg. 63

V.

THE ARGUMENT FROM EMBRYOLOGY.

There is still another important line of evidence which we cannot afford to overlook; I mean the argument from embryology. To economise space, I shall not explain the considerations which obviously lead to the anticipation that, if the theory of descent by inheritance is true, the life history of the individual ought to constitute a sort of condensed epitome of the whole history of its descent. But taking this anticipation for granted, as it is fully realised by the facts of embryology, it follows that the pg. 64science of embryology affords perhaps the strongest of all the strong arguments in favour of evolution. From the nature of the case, however, the evidence under this head requires special training to appreciate; so I will merely observe, in general terms, that the higher animals almost invariably pass through the same embryological stages as the lower ones, up to the time when the higher animal begins to assume its higher characters. Thus, for instance, to take the case of the highest animal, man, his development begins from a speck of living matter similar to that from which the development of a plant begins. And, when his animality becomes established, he exhibits the fundamental anatomical qualities which characterise such lowly animals as the jelly-fish. Next he is marked off as a vertebrate, but it cannot be said whether he is to be a fish, a snake, a bird or a beast. Later on it is evident that he is to be a mammal; butpg. 65 not till still later can it be said to which order of mammals he belongs.

Now this progressive inheritance by higher types of embryological characters common to lower types is a fact which tells greatly in favour of the theory of descent, whilst it seems almost fatal to the theory of design. For instance, to take a specific case, Mr. Lewes remarks of a species of salamander—which differs from most salamanders in being exclusively terrestrial—that although its young ones can never require gills, yet on cutting open a pregnant female we find the young ones to possess gills like aquatic salamanders; and when placed in the water the young ones swim about like the tadpoles of the water newt. Now, to suppose that these utterly useless gills were specially designed is to suppose design without any assignable purpose; for even the far-fetched assumption that a unity of ideal is the causepg. 66 of organic affinities, becomes positively ridiculous when applied to the case of embryonic structures, which are destined to disappear before the animal is born. Who, for instance, would have the courage to affirm that the Deity had any such motive in providing, not only the unborn young of specially created salamanders, but also the unborn young of specially created man, with the essential anatomical features of gills?

But this remark leads us to consider a little more attentively the anatomical features presented by the human embryo. The gill-slits just mentioned occur on each side of the neck, and to them the arteries run in branching arches, as in a fish. This, in fact, is the stage through which the branchiæ of a fish are developed, and therefore in fish the slits remain open during life, while the so called "visceral arches" throw out filaments which receive the arterial branches coming from the aortic arches, and so become pg. 67the organs of respiration, or branchiæ. But in all the other vertebrata (*i.e.* except fish and amphibia) the gill-slits do not develop branchiæ, become closed (with the frequent exception of the first), and so never subserve the function of respiration. Or, as Mr. Darwin states it, "At this period the arteries run in arch-like branches, as if to carry the blood to branchiæ which are not present in the higher vertebrata, though the slits on the sides of the neck still remain, marking their former position."

The heart is at first a simple pulsating vessel, like the heart of the lowest fishes, and the excreta are voided through a common cloacal passage — an anatomical feature so characteristic of the lower vertebrata, that it occurs in no fully formed member of the mammalian group, with the exception of the bird-like order of monotremata, which takes its name from presenting so striking a peculiarity.pg. 68

At a later period the human embryo is provided with a very conspicuous tail, which is considerably longer than the rudimentary legs occurring at that period of development, and which Professor Turner has found to be provided with muscles — the extensor, which is so largely developed in many animals, being especially well marked.

Again, as Mr. Darwin says, "In the embryos of all air-breathing vertebrates, certain glands, called the corpora Wolffiana, correspond with and act like the kidneys of mature fishes;" and during the sixth month the whole body is covered very thickly with wool-like hair—even the forehead and ears being closely coated; but it is, as Mr. Darwin observes, "a significant fact that the palms of the hands and the soles of the feet are quite naked, like the inferior surfaces of all four extremities in most of the lower animals," including monkeys.

Lastly, Professor Wyman has found that in apg. 69 human embryo about an inch in length, "the great toe was shorter than the others; and, instead of being parallel to them, projected at an angle from the side of the foot, thus corresponding with the permanent condition of this part in the quadrumana."[1]

Therefore, on the whole, we may conclude these brief remarks on embryology with the words of Professor Huxley:—"Without question, the mode of origin, and the early stages of the development of man, are identical with those of the animals immediately below him in the scale; without a doubt, in these respects he is far nearer to apes than the apes are to the dog."[2]

[1]*Proc. Amer. Acad. Scs.*, vol. iv., 1860, p. 17. It should be added, however, that although the direction taken by the great toe of man at this early age is doubtless, as Prof. Wyman states, more like that which obtains in the quadrumana, there is a slight anatomical difference in the mode of its articulation with the foot, which seems to assist in securing the forward direction taken by it in later life.

[2]*Man's Place in Nature*, p. 65.

pg. 70

VI.

**ARGUMENTS DRAWN FROM CERTAIN GENERAL CONSID-
ERATIONS.**

There are two or three arguments of a somewhat weighty character, which do not fall under any of the previous headings, but which
we must not on this account neglect.

1. It is justly deemed a substantiation of a scientific theory if it is
found to furnish an explanation of other classes of phenomena than
those for the explanation of which it was first devised. And this is
the case with the theory of natural selection in the region of psychology. The theory was first devised to explain thepg. 71 facts of
biology, and proving so successful in that region, Mr. Darwin pro-
ceeded to test it in the region of psychology. The result has been to
show that large classes of phenomena in this region which were
previously unaccountable become fully intelligible. This is especially the case with the phenomena of instinct, and in a lesser degree
with those of reason and conscience. For the theory shows that if
structures admit of being moulded to their special uses by natural
selection, the same must be true of instincts; and it is found an easy
matter to understand how, by seizing upon and fixing, through
hereditary beneficial variations of habit (whether instinctive or intelligent), natural selection is as competent to fashion the mental
structure of an animal as it is to shape its bodily structure into
agreement with the external conditions of life. Thus the whole philosophy of animal intelligence is greatly elucidated,pg. 72 and this
fact may justly be regarded as lending much additional credence to
the theory.

Again, by observing that sympathy and the social instincts generally are developed to a large extent in many of the lower animals,
and particularly so in the quadrumana, the theory of natural selection is provided with a reasonable basis for furnishing a scientific
explanation of the moral sense in man; and by observing that many
of the lower animals are capable of drawing simple inferences, the
theory is likewise able to explain the development of reason. So that
in the province of human psychology no less than in that of animal,
the theory of natural selection, in showing itself competent to ex-

plain much which is otherwise inexplicable, is seen to derive a large additional measure of argumentative support.

2. Although the majority of structures and instincts met with in the animal kingdom arepg. 73 in a marvellous degree suited to the performance of their functions and uses, it is nevertheless far from being an invariable rule that the suitability is perfect. Thus, for instance, even in the case of the eye — which is perhaps the most wonderful and most highly elaborated structure in organic nature — it is demonstrable that the organ, considered as an optical instrument, is not ideally perfect; so that, if it were an artificial production, opticians would know how to improve it. And as for instinct, numberless cases might be adduced of imperfection, ranging in all degrees from a slight deficiency to fatal blundering.

Now if all organic structures are supposed to be mechanisms designed by the Deity, and all instincts are supposed to be mental attributes implanted by Him, it becomes unintelligible that in the result the human mind should thus be able to perceive, either anpg. 74 ignorance of natural principles in the Author of nature, or a singular absence of thought in applying His knowledge. But, on the other hand, if all the structures and instincts are supposed to be due to natural selection (whether alone or in conjunction with other natural causes), we have no need to feel staggered at flagrant cases of imperfection; we have only to wonder at the number of cases in which perfection, more or less complete, has been attained.

3. Lastly, there is still another general consideration, and one which appeals to my mind as of immense weight. The question, it will be remembered, lies between beneficent design and natural selection, and I think that the consideration about to be adduced is in itself alone sufficient to decide the question.

This consideration is that amid all the millions of mechanisms and instincts in the animal kingpg. 75dom, there is no one instance of a mechanism or instinct occurring in one species for the exclusive benefit of another species, although there are a few cases in which a mechanism or instinct that is of benefit to its possessor has come also to be utilised by other species. Now, on the beneficent design theory it is impossible to explain why, when all the mechanisms in the same species are invariably correlated for the benefit of that species, there should never be any such correlation between mechanisms in different species, or why the same remark should apply to instincts. For how magnificent a display of divine beneficence would organic nature have afforded, if all, or even some, species had been so inter-related as to minister to each other's necessities. Organic species might then have been likened to a countless multitude of voices all singing in one harmonious psalm of praise. But, as it is, we see no vestige of such co-ordination;pg. 76 every species is for itself, and for itself alone—an outcome of the always and everywhere fiercely raging struggle for life.

Such, then, is a sketch of the evidence in favour of organic evolution. Of course in such a meagre outline it has not been possible to do justice to that evidence, which should be studied in detail rather than looked at in such a bird's-eye view as I have presented. Nevertheless, enough, I hope, has been said to convince all reasonable persons, that any longer to withhold assent from so vast a body of evidence is a token, not of intellectual prudence, but of intellectual incapacity. With Professor Huxley, therefore, I exclaim,—"Choose your hypothesis; I have chosen mine," and "I refuse to run the risk of insulting any sane man by supposing that he seriously holds such a notion" as that of special creation. These words, I submit, are not in the least too strong; for if any manpg. 77 can study the many and important lines of evidence all converging on the central truth that evolution has been the law of organic nature, and still fail to perceive the certainty of that truth, then I say that that man—either on account of his prejudices, or from his inability to estimate the value of evidence—must properly be regarded as a weak-minded man. Or, to state the case in another way, if such a man were to say to me,—Notwithstanding all your lines of evidence, I still believe in special design manifested in creation; I should reply,—And in this I fully agree with you; for if, notwithstanding these numerous and important lines of evidence, the theory which they substantiate is false, then to my mind we have the best conceivable evidence of very special design having been manifested in creation—the special design, namely, to deceive mankind by an elaborate, detailed, and systematic fraud. For, if thepg. 78 theory of special creation is true, I hold that as no one fact can be adduced in its favour, whilst so vast a body of facts can be adduced against it, the only possible explanation of so extraordinary a circumstance is that of a mendacious intelligence of superhuman power carefully disposing all the observable facts of his creation in such a way as to compel his rational creatures, by the best and most impartial use of their rational faculties, to conclude that the theory of evolution is as certainly true as the theory of special creation is conspicuously false.

But having now concluded this brief review of the leading arguments in favour of organic evolution, and having expressed as forcibly as I am able my own opinion upon them, I do not wish it to be supposed, either that I am intolerant of opinions which are held by others, or that I have been trying to, "make out apg. 79 case" by suppressing adverse facts. I am not intolerant, because I believe that dissent from the general doctrine of evolution can only arise either from ignorance of some special departments of science, or from a bias of feeling against the doctrine—to both of which weaknesses evolutionists can afford to be indulgent. And in order to show that I have not been trying unfairly to make out a case, I shall conclude by briefly reviewing the arguments which have been adduced against the doctrine in question.

The only argument of this kind that I know from the side of reason (if we neglect those special objections which have been fully shown by Mr. Darwin himself to be based on inadequate information or erroneous conception, and therefore futile), is that which says:—Evolution, if true, can only be proved so by an actual observation of the process, and as no one pretendspg. 80 to have witnessed the transmutation of species, it follows that evolution has not been proved.

Now, it is perfectly right to draw a clear distinction between a theory and a demonstration; but it is a great mistake to suppose that a theory may then only be admitted by science when it has been demonstrated. Bishop Butler tells us that "Probability is the guide of life," and not less true is it that probability is likewise the guide of science. The business of science, as of common life, is to estimate correctly the relative degrees of probability presented by this and that theory or hypothesis; when once a theory or hypothesis is demonstrated it ceases to be a matter of scientific inquiry, and becomes a matter of scientific fact. Thus received, we have to consider the doctrine of evolution as certainly standing in the first rank of scientific theories in respect of probability sustained by evidence, although no lesspg. 81 certainly not demonstrated as a matter of scientific fact. But when a theory has been raised to such a level of probability as this, it is, for all practical purposes, as good as a demonstration. Thus, in the particular instance before us, even if the sceptical demand for evidence, which from the nature of the case is clearly impossible, were granted, and if we could actually observe the transmutation of species, the fact would not exert any further influence on the progress of science than is now exerted by the large and converging bodies of evidence which leave no other rational theory open to us than that such transmutation has taken place. Therefore, it seems to me, the hypercritical objection which we are considering is really founded on a misconception of scientific method, and of what it is that justifies a scientific doctrine. Assuredly, in the case of every theory, as distinguished from a demonstration, there must always be apg. 82 proportion between the evidence of and the warrant for the proposition which the theory states; and if gauged by this simple rule the warrant for accepting the theory of evolution is now estimated by the judgment of all scientifically trained minds as so high, that by no additional evidence could it be placed higher without becoming a full demonstration. Or, otherwise stated, as a theory the doctrine of descent is now in the topmost position of probability, so that by no amount of additional evidence could it be raised higher without ceasing to be a probability and becoming a certainty. That is to say, we do not need any more evidence in any of the lines of evidence to add to the strength of our belief in, as distinguished from our knowledge of, the truth of evolution. For the strength of our conviction could not be increased by the discovery of any additional number of connecting links amongpg. 83 fossil species, further facts relating to geographical distribution, to morphology, classification, embryology, or any of the other lines of evidence which have been mentioned; no further evidence the same in kind is now competent to raise in degree the probability which has already been raised, as far as from its very nature as a probability it can be raised.

I have no doubt, however, that the principal obstacle which the doctrine of evolution encounters in the popular mind is not one of reason, but of sentiment. It is thought that the conception of man being a lineal descendant of the monkey is a conception which is degrading to the dignity of the former animal. Now this obstacle being, as I have said, a matter of feeling or sentiment, as such I am not able to meet it. If you think that man is shown to be any less human because his origin is now shown to have been derivative, I cannot change that decision on your part; Ipg. 84 can only express dissent from it on my own. But although I cannot affect your sentiments in this matter, I may be permitted to point out that, as they are only sentiments, they are quite worthless as arguments or guides to truth. I have yet to learn that the "dignity of man" is a matter of any concern to our Mother Nature, who in all her dealings appears, to say the least, to treat us in rather a matter-of-fact sort of way. Indeed, so far is she from respecting our ideas of "dignity," that whenever these ideas have been applied to any of her processes, the progress of science has been destined rudely to dispel them. Thus, for instance, when the sun-spots were first observed they were indignantly denied by the Aristotelians, on the ground of its being "impossible that the eye of the universe could suffer from ophthalmia;" and when Kepler made his great discovery of the accelerated and retarded motion of the planets in different parts of their orbits, many personspg. 85 refused to entertain the conception, on the ground that it was "undignified" for heavenly bodies to hurry and slacken their pace in accordance with Kepler's law. This now seems most absurd to us; but to posterity it will not seem nearly so much so as that, notwithstanding such precedents, persons should still be found to object to Darwin's discovery, not because they were anxious to maintain the dignity of the heavenly bodies, but because they were so ludicrously anxious to maintain the dignity of their own! Good it is for man, puffed up with such silly pride, that Nature teaches him humility.

But, before leaving this subject, I should like further to point out that those who advance this preposterous objection from dignity appear to forget one all-important point, viz., that whether or not the monkey is the parent of the man, the man is certainly made in every way to _look like_ a child of the monkey. For it is a matterpg. 86 of anatomical demonstration, that in all the features of our bodily structure—even up to our brains—we more closely resemble the man-like apes than the man-like apes resemble the lower quadru-mana. And I beg it to be remembered that the tremendous signifi-cance of this fact can only be duly appreciated by those who know the astounding complexity of our bodily structure. Those who are ignorant of human anatomy cannot form any adequate — probably not even an approximate — conception of its intricacy. Yet we find that this terrifically intricate organisation is repeated down to all the minute bones and muscles, blood-vessels, nerves and viscera, in the bodies of the higher apes. Here, then, I say, we have a fact — or ra-ther let me say a hundred thousand facts — which cannot possibly be attributed to chance. As reasonable beings we must conclude that there has been some definite cause for this extraordinary imitation by the mostpg. 87 highly organised being in creation of the next most highly organised. And if we reject the natural explanation of hereditary descent from a common ancestry, we can only suppose that the Deity, in creating man, took the most scrupulous pains to make him in the image of the ape. This, I say, is a matter of undeni-able fact — supposing the creation theory true — and as a matter of fact, therefore, it calls for explanation. Why should God have thus conditioned man as an elaborate copy of the ape, when we know from the rest of creation how endless are His resources in the inven-tion of types?

I present the matter thus to show that even the weight of senti-ment is not all on the side of special creation. Look on this picture and on this: —

The Creator has exhibited the extraordinary and unaccountable design of casting the complexpg. 88 structure of man in the same mould that He had just previously used to cast the complex struc-ture of the ape.

"When I view all beings, not as special creations, but as the lineal descendants of some few beings which lived long before the first bed of the Cambrian system was deposited, they seem to me to become ennobled.... There is grandeur in this view of life, with its several powers, having been originally breathed by the Creator into a few forms or into one; and that, whilst this planet has gone cycling on according to the first law of gravity, from so simple a beginning endless forms most beautiful and most wonderful have been and are being evolved."

THE END.

NATURE:

A Weekly Illustrated Journal of Science.

NATURE contains Original Articles on all subjects coming within the domain of Science, contributed by the most eminent Scientists, belonging to all parts of the world.

Reviews, setting forth the nature and value of recent Scientific works, are written for Nature by men who are acknowledged masters in their particular departments.

The Correspondence columns of Nature, while forming a medium of Scientific discussion and of intercommunication among the most distinguished men of Science, have become the recognized organ for announcing new discoveries and new illustrations of Scientific principles among observers of Nature all the world over—from Japan to San Francisco, from New Zealand to Iceland.

The Serial columns of Nature contain the gist of the most important Papers that appear in the numerous Scientific Journals which are now published at home and abroad, in various languages; while longer Abstracts are given of the more valuable Papers which appear in foreign Journals.

The Principal Scientific Societies and Academies of the world, British and foreign, have their transactions regularly recorded in Nature, the Editor being in correspondence, for this purpose, with representatives of Societies in all parts of the world.

Notes from the most trustworthy sources appear each week recording the latest gossip of the Scientific world at home and abroad.

As questions of Science compass all limits of nationality, and are of universal interest, a periodical devoted to them may fitly appeal to the intelligent classes in all countries where its language is read. The proprietors of Nature aim so to conduct it that it shall have a common claim upon all English-speaking peoples. Its articles are brief and condensed, and are thus suited to the circumstances of an active and busy people who have little time to read extended and elaborate treatises.

Subscriptions to "Nature":

Yearly 28*s.* | Half-Yearly 14*s.* 6*d.* Quarterly 7*s.* 6*d.*

To the Colonies, United States, the Continent, and all places within the Postal Union:

Yearly 30*s.* 6*d.* | Half-Yearly 15*s.* 6*d.* Quarterly 8*s.*

P.O.O. to be made payable to MACMILLAN AND CO.

Office: 29 BEDFORD STREET, STRAND